YOUR KNOWLEDGE HAS

- We will publish your bachelor's and master's thesis, essays and papers

- Your own eBook and book - sold worldwide in all relevant shops

- Earn money with each sale

Upload your text at www.GRIN.com and publish for free

Bibliographic information published by the German National Library:

The German National Library lists this publication in the National Bibliography; detailed bibliographic data are available on the Internet at http://dnb.dnb.de .

This book is copyright material and must not be copied, reproduced, transferred, distributed, leased, licensed or publicly performed or used in any way except as specifically permitted in writing by the publishers, as allowed under the terms and conditions under which it was purchased or as strictly permitted by applicable copyright law. Any unauthorized distribution or use of this text may be a direct infringement of the author s and publisher s rights and those responsible may be liable in law accordingly.

Imprint:

Copyright © 2014 GRIN Verlag, Open Publishing GmbH
Print and binding: Books on Demand GmbH, Norderstedt Germany
ISBN: 9783668607415

This book at GRIN:

https://www.grin.com/document/386648

Robert J. G. Wenndorff

Simulation and Optimization of the Currency in a Matlab Model. Visualization of the Currency and the Voltage in Dependency of the Anchors Force

GRIN - Your knowledge has value

Since its foundation in 1998, GRIN has specialized in publishing academic texts by students, college teachers and other academics as e-book and printed book. The website www.grin.com is an ideal platform for presenting term papers, final papers, scientific essays, dissertations and specialist books.

Visit us on the internet:

http://www.grin.com/

http://www.facebook.com/grincom

http://www.twitter.com/grin_com

HOCHSCHULE PFORZHEIM

Simulation of the currency and voltage in a MATLAB Simulink model

This paper is based on a MATLAB Simulink model showing an anchor movement in a direct current magnet. The result is to get a visualization of the currency I in [A] and the voltage U in [V] in dependency of the anchors force.

Project worker

Robert J. G. Wenndorff

Executive Summary

In the "Konstruktionsseminar Summer term 2014" the idea is to optimize a given artificial heart by the last semester. The target is to reduce the weight and to raise the force of the heart engine. Some constructive ideas took place, for example the material has been changed from machining steel to an iron-cobalt material to have a higher flux density. For the engine, a solenoid is used to give a linear movement of the anchor to push blood into each heart chamber. In the high-pressure side, the coil will be set under currency to produce magnetic force via Ørsted's law.

The system will be seen as a mass-spring-damper-system and will be analyzed in Matlab Simulink. With FEMM, a tool to simulate finite elements method for magnetism, the forces and inductivities, the airgap and the currency of the solenoid can be taken and implemented into Matlab Simulink. The target is to control the input of the designed system with a change of the voltage level, so that the currency can be increased and decreased and the movement of the anchor can be controlled over the whole airgap.

Due having an idly system, the voltage has to be regulated into an extremely short period that the time for the anchor can be set for half a second. This period of time is necessary to have a heart rate of about 60 beats per minutes. During the other half second, a spring pushed the anchor back into the initial condition and the cycle can start again. Having a shorter period of time means that the blood as a fluid doesn't move permanently and a result could be the bursting of veins in the body where the resistance is low. Or for example a turbulent flow can be produced which block veins in thin regions, so there will be a blood jam and parts of the body cannot receive oxygens.

After putting all incoming facts into the Matlab Simulink model, the simulation shows that the time for the anchor to move in the airgap is too short with about $0,1s$ that all those negative facts may come true.

Having a system like this, the control of the voltage can be made manually but the idly of the dependent currency makes it hard to control the movement:
- The force to push the blood cannot be reached
- The time for the anchor to move is too short
- The spring cannot push the anchor back due the idly of the currency

List of Contents

Project worker i
Supervisor at Pforzheim University of applied sciences i
Executive Summary ii
List of Contents iii
Glossary iv
Formula Table v
Formula Symbols with Latin Characters v
Formula Symbols with Greek Characters v
Indexes v

1	**Introduction**	**1**
2	**Description of the task**	**2**
3	**Initial situation**	**3**
4	**State of the Art**	**4**
4.1	Magnetic force FM	5
4.2	Duty cycle	5
4.3	Inner resistance of the coil	6
4.4	Derivation of MSDS – Mass-Spring-Damper-System	8
4.4.1	The spring stiffness	8
4.4.2	The damper stiffness	10
4.4.3	The anchors mass	10
4.5	Airgap	10
5	**From FEMM into a Simulink model**	**11**
5.1	FEMM	11
5.2	Creating the Matlab Simulink model	13
5.2.1	Electric part of the Simulink model	13
5.2.2	Mechanical part of the Simulink model	14
6	**Use of the system**	**15**
7	**Conclusion**	**16**
8	**Table of Figures**	**18**
9	**References**	**19**
11	**Appendix**	**20**
11.1	FEMM Files	20
11.1.1	F(delta, I)	20
11.1.2	F(delta, L)	20
11.2	Coil Calculations	21

Glossary

bpm: bpm is a short form of "beats per minute". This is the heart beat of a human adult without any kind of sportive efforts.

bps: bps is a short form of "beats per second". It is bpm divided by 60 to get the beats per second. $1 bpm = 60 bps$.

DC: DC is the short form for "duty cycle". Duty cycle is the particular time when the voltage on.

FEMM: FEMM is the acronym for Finite Element Method Magnetics. It is a numeric FE program to solve stationary static magnetic field problems in a two-dimensional way.

HP-side: HP-side stands for the high-pressure side. This is the active chamber of the artificial heart. It connects the heart with the body blood loop.

LP-side: LP-side stands for the low-pressure side. This is the passive chamber of the artificial heart. It connects the heart with the lung blood loop.

MSDS: MSDS is a short form for mass-spring-damper system, which gives us the equation of motion.

PT1: PT1 is delay element or lag element in the control engineering. This element is a continuous rising element until it reaches its upper limit.

Formula Table

Formula Symbols with Latin Characters

Symbol	Unit	Relevance
a	m/s^2	Acceleration.
c	N/m	Spring stiffness.
d	$N*s/m$	Damper stiffness.
F	N	Force.
I	A	Currency.
l	mm	Length of the Wire.
L	$V*s/A$	Inductivity.
m	kg	Mass.
n		Windings of the coil.
r	mm	Height Coil in the blue or red Area.
R	Ω	Resistance.
s	m	Airgap.
t	s	Time.
U	V	Voltage.
v	m/s	Velocity.
z	mm	Breadth Coil in the blue or red Area.

Formula Symbols with Greek Characters

Symbol	Unit	Relevance
α	K^{-1}	Temperature Coefficient.
ϑ	K	Max. Temperature.

Indexes

Symbol	Relevance
0	Condition at point 0. Initail condition.
1	Condition at point 1. Condition after condition 0. Number of Countings.

2	Condition at point 2. Condition after condition 1. Number of Countings.
a	Outer Radius. For example, the Coil.
$b; r$	Blue or red Area.
curr	Current.
d	Damper.
HP	High pressure.
i	Inner Radius. For example, the Coil.
LP	Low pressure.
M	Magnet, Solenoid.
$m_b; m_r$	Middle Radius of the Coil in the blue or red Area.
r	Windings in r-Direction in the blue or red Area.
s	Spring. Radius Coil, Cant.
z	Windings in z-Direction in the blue or red Area.

1 Introduction

The present work is about the design of a linear magnet for an artificial heart in the "Konstruktionsseminar Summer term 2014" at the University of Applied Science HS Pforzheim. This work is one part of three different topics about the heart in the course "electric machines". It covers the simulation and optimization of the engine from an artificial heart using a linear magnet or further named solenoid. In the winter term 2013, the last semester, a group of students already designed a virtual prototype of the heart and the magnet. But they could not meet the weight requirements therefore the goal and the focus in the "Konstruktionsseminar Summer term 2014" was to lower the total weight. However, no analyses had been made of the movement of the anchor and pushing the blood into the chamber. To lower the weight and to raise the power of the solenoid by an exchange of the material, the whole system will be more dynamic as before. To get an optimum of a low weight plus having a dynamic system, which can be controlled by changing the voltage and therefore the currency, the system is getting into an idly condition.

2 Description of the task

Given is an artificial heart, using a solenoid as kind of an engine for the engineered product. The magnet should give the patient an opinion for a normal life.
The basic idea is using a solenoid like a common rail injection pump as an actuator in the system. The solenoid is placed centered in the artificial heart moving itself in a horizontal direction. On each side is a chamber with an income and outgoing flow of blood. The magnet is connected to a membrane on each side, minimizing the chambers space and to get the blood flow. To move the solenoid's anchor, a specific currency and voltage has to be connected via a battery. There is only one side the anchor is moving to, for the other, there is spring which pushes the anchor back in the opposite direction. So, there is a high- and a low-pressure side. The high-pressure side is in the active anchor direction, where the magnet force is pointing to, to have the higher force in just that chamber. This side pushed the blood into the body loop. The spring pushed the anchor back in the low-pressure side, the lung loop.

The voltage is changing from off to on and the currency starts rising. An electric field is available and the anchor is moving due to Ørsteds law.[1]

This represents a mass-spring-damper system. The moving mass is the anchor, the spring is placed to push the anchor back in the starting condition and the damper is given to minimize the impact of the anchor to each ground side plus the blood damper while with the anchor pushed the membrane into the blood in the chamber and continuing the blood in the body loop.

The MSDS can be built up in a MATLAB Simulink model to simulate the movement of the anchor with a defined currency. Due to the fact this solenoid should be implemented into a human body the temperature is a big problem. Therefore, voltage and currency are imitating factors for the solenoid. The purpose is to get the currency for the solenoid to move it from the low to the high-pressure side until the impact with the ground side.

---

[1] Heidrich (2014): Magnetism; S.23f

3 Initial situation

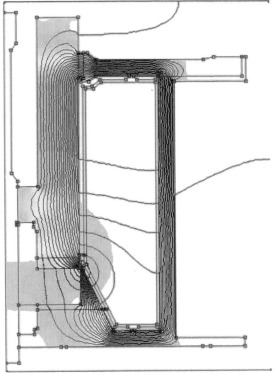

Figure 1 – Final situation of the solenoid from the group of winter term 13/14[2]

An optimized geometric design from a previous group is given in Figure 1. Major changes, seen in Figure 2, to the previous groups are the exchange of the anchor to an iron-cobalt material. Additionally, the geometry has been changed to minimize volume and weight. Radiuses are added to fulfill a good and direct current flow.

---

[2] Riewe et al. (2013): Vollständige Nachbildung des menschlichen Herzens

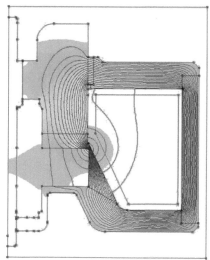

Figure 2 - Initial situation at the beginning of the simulation and optimization[3]

There have no previous tasks like this, to control the currency and voltage in a MATLAB/Simulink model, taken place before.

4 State of the Art

In this section, every part of the system will be explained and calculated manually. First, the magnetic force which is mandatory to move the anchor. More information will be in chapter 4.1. Second, the duty cycle: a moving magnet is not a straight line by moving from one to another position. So, this will be explained in chapter 4.2. In chapter 4.3 the inner resistance of the coil and the length of the copper wire are taken place. Next part is an introduction of the MSDS. This also includes the spring and damper stiffness and the anchor mass. For more info go to chapter 4.4. The last chapter, 4.5, is about the airgap, which is a variable over the whole process of movement.

---

[3] Rommelfanger et al. (2014): Nachbildung des menschlichen Herzens

4.1 Magnetic force F_M

The force of the solenoid, which pushes into the high-pressure side, must be as high as the area A_{HP} times the pressure on the high-pressure side p_{HP}. It also has to push the pre-stressed spring against the working force F_S.[4]

$$F_M = p_{HP} * A_{HP} + F_S \qquad \text{equation (1)}$$

Pressure p_{HP} and area A_{HP} are calculated by using values of an adult person. Normally one push into the high-pressure side is about $p_{HP} = 170 \, mbar = 1{,}7 * 10^4 \, \frac{N}{m^2}$. The area A_{HP} is the size of the membrane which pushed the blood into the body loop. This value can be calculated by knowing that every minute about 4 liter of blood has to be pumped through the human body without any sportive efforts.[5] With a heart beat of about 60 bpm, every beat has to push about

$$\frac{4000 \, mliter}{1 \, bpm * 60} = 67 \, \frac{ml}{s} \qquad \text{equation (2)}$$

67 milliliter per second into the human body. This value has to divided by the chambers height to get the area, which is about $4000 mm^2$.[6]

$$F_M = 1{,}7 * 10^4 \, \frac{N}{10^6 mm^2} * 4000 mm^2 + 25N = 95N \qquad \text{equation (3)}$$

The solenoid has to produce a force F_M of about 95N constantly.

4.2 Duty cycle

Applying the voltage to a higher value, the currency will also increase by time. In this coil, voltage at the start is set from $U = 0V$ to a higher value. In a function diagram, it is a step response.
The duty cycle can be calculated with the control engineering mathematics. The duty cycle of a solenoid is basically a PT1-element giving a step response. The increasing ramp of this unit step response can be calculated with the following equation:[7]

$$Y_{PT1} = K_{PT1} * \left(1 - e^{-\frac{t}{T_{PT1}}}\right) \qquad \text{equation (4)}$$

---

[4] Kallenbach et al. (2012): Elektromagnete; S.295f
[5] Bauer and Enneker (2008): Herzklappenchirurgie; S.10
[6] Calculation in Rommelfanger et al. (2014): Nachbildung des menschlichen Herzens
[7] Boege and Boege (2013): Handbuch Maschinenbau; S.H17

The following figure is a symbol figure for such a unit step response:

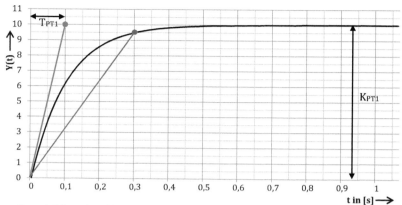

Figure 3: symbol figure for a PT1 unit step response[8]

K_{PT1} is the currency, the function is heading to. T_{PT1} is the time constant, describing the increase of the function. The tangent in the red line, of the starting point, hits the final value at 100%. At 95% of the function can be calculated with:

$$0.95 * K_{PT1} = Y_{PT1}(3 * T_{PT1}) \qquad \text{equation (5)}$$

The voltage for the duty cycle will just applied for some particular seconds. It will be decreased because of the magnetic force and the air gap which is getting smaller. It will not directly be turned off; the reduction will be stepwise to get a continually movement of the

4.3 Inner resistance of the coil

The inner resistance of the coil can be determined with the geometry and the resistivity of the copper wire. For a temperature of 20°C, it can be calculated with the formula:[9]

$$R_{20} = \rho * \frac{L}{A} \qquad \text{equation (6)}$$

---

[8] Self creation in Matlab Simulink
[9] Heidrich, Peter (2014): Magnetism; S.8

The resistivity of the copper wire is approximately $0.0175\,\Omega * \frac{mm^2}{m}$. The cross-sectional area A is determined by the diameter of the wire.

$$A = \frac{\pi}{4} * d^2 = \frac{\pi}{4} * (0{,}63mm)^2 = 0{,}312\ mm^2 \qquad \text{equation (7)}$$

The length of the copper wire can be calculated by the geometry of the coil. The cant of the coil gives a split of the coil in two areas, a red and a blue one. Furthermore, the number of windings has be defined in every area. Those measurements are taken from the CAD model.

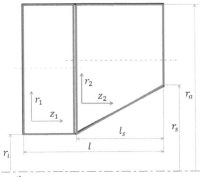

Figure 4 - Measurements of the coil

l	$20mm$
l_s	$12{,}85mm$
r_a	$30mm$
r_s	$20{,}9mm$
r_i	$15{,}4mm$
d	$0{,}63mm$

The final length is set with $91371mm$. The calculation of the length is in 11.2 "Coil Calculations".

From equation (6), the results for the inner resistance is $R_{20} = 5.125\,\Omega$.
As the inner resistance is a temperature-dependent variable, the calculation is by using the max. temperature of the coil in action. According to simulations with an FEM-program, of the thermo-group, the maximal temperature on the coil amounts approximately $50\,°C$. The dependency of the resistance is described by following equation:

$$R_\vartheta = R_{20} * (1 + \alpha * (\vartheta - 293K)) \qquad \text{equation (8)}$$

The temperature coefficient α of cupper is $3.93 * 10^{-3} K^{-1}$. ϑ describes the max. temperature of 50 °C, which is equal to $\vartheta = 323K$. From equation (8), the result for the inner resistance is R_{50} = 5.729 Ω.

4.4 Derivation of MSDS – Mass-Spring-Damper-System

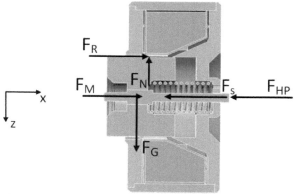

Figure 5 – MSDS of the solenoid

A MSDS is a system in which the forces of the spring, damper and the mass of the moving part are shown. In general, it is a sum of all the components derived of the chapters 4.4.1, 4.4.2 and 4.4.3. Depending of there is an external force pushing on the system the equation is zero or the external force. In this case the MSDS is a mass oscillator with a single degree of freedom.[10]

$$F_{ext} = m * a + d * v + c * s \qquad \text{equation (9)}$$

4.4.1 The spring stiffness

Mostly a spring is produced by a high-quality steal and with the roundings a spring has a kind of stiffness. This stiffness is important that the spring can be compressed and uncompresses itself to its origin length. The spring is axially loaded with the force F_M produced by the solenoid. Within time the spring compresses over the Hooke's law until the airgap, chapter 4.5, is 0. To calculate the spring stiffness c, which is used in the Mablab Simulink model, the following equation is used:[11]

---
[10] Boege and Boege (2013): Handbuch Maschinenbau; S.H18f
[11] Boege and Boege (2013): Handbuch Maschinenbau; S.H18

$$F_s = c * s \qquad \text{equation (10)}$$

F_s is the force the spring pushed in the opposite direction, c is the spring stiffness and s the way the spring is getting pressed in the axial direction. The spring is being prestressed in the system to have always a kind of stress in the MSDS. To calculate the stiffness c in a presteressed system the equation (10) will have a difference between the upper and the lower value, delta.

$$c = \frac{\Delta F_s}{\Delta s} = \frac{F_1 - F_0}{s_1 - s_0} \qquad \text{equation (11)}$$

The force F_s has to be as high as the area of low pressure needs the force. Therefore, the calculation of c is:

$$c = \frac{15{,}5791 - 15{,}7190}{0{,}0179 - 0{,}011} = 1429 \, \frac{N}{m} \qquad \text{equation (12)}$$

The unit of c is $\frac{N}{m}$. This value is implemented in the Matlab Simulink model.

Because $F_M > F_s$ the spring compresses until the electric field decreases and $F_M \leq F_s$. At this moment, the spring pushes the anchor back to the initial situation. The energy which is saved in the system as a potential energy will be unlocked as kinetic energy.

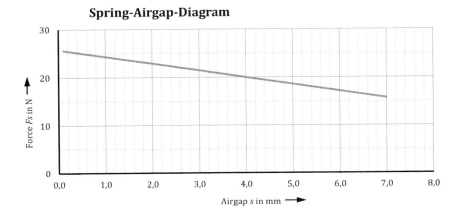

4.4.2 The damper stiffness

The calculation of the damping stiffness depends on the viscosity of the blood and the inertia of the membrane which has to be pushed. Both of those simulations have to take place via FEM. The resistance force produced by the high-pressure side is the same as F_d. With the equation (13) the force F_d, which is the result of damper stiffness d and the velocity of the anchor, moving the blood and the membrane, v.[12]

$$F_{HP} = F_d = d * v \qquad \text{equation (13)}$$

Due there was no value available; the damper stiffness is approximately set as $0{,}4 \ \frac{N*s}{m}$.

In a spring-mass-pendulum with a damping stiffness of zero, the function would be sinus curve. With the usage of a damper the spring-mass-pendulum would stop swinging. It is the aim to set the damper stiffness, which works against the forces of F_M.

4.4.3 The anchors mass

The mass of the anchor is taken from the CAD model. The construction group set the mass with $m = 0{,}1 kg$. This value is implemented in the Matlab Simulink model.
The force F_M accelerates the mass in the side of the high pressure. Due the mass is given, the equation (14) shows the acceleration of this mass.[13]

$$F_M = m * a \qquad \text{equation (14)}$$

4.5 Airgap

The airgap is given by the system. It is the difference of maximum length, when the solenoid in not in an electric field, and the minimum length, when the anchor contacts the housing to have the optimum of the electric field.
During the construction and after calculations of the volume the solenoid has to push into its magnetic direction, the airgap is given with $s = 0{,}007m$. This airgap is needed to move the membrane. To minimize the destruction of the membrane, the area of the membrane has been maximized to a big round area to optimize the volume of blood being moved in each heartbeat.

---

[12] Boege and Boege (2013): Handbuch Maschinenbau; S.H18
[13] Boege and Boege (2013): Handbuch Maschinenbau; S.H18

5 From FEMM into a Simulink model

Via the tool FEMM4.2, a tool solving magnetic static stationary problems, the movement of the solenoid can be simulated. To get a dynamic model, every step of the anchors movement has to be simulated. Not just that to get a qualitatively response of the solenoid, every movement step is getting simulated over various currency steps. So, all in all, there are two variables to simulate the MSDS model:
- Currency I in [A]
- Airgap s in [m]

The solenoid is simulated in every position, seen in 11.1.1 F(delta, I) and 11.1.2 F(delta, L). The results can be implemented into the MSDS.

Another tool, Matlab Simulink, is used to design and control the loop of the MSDS. The exported values of FEMM can be integrated into this tool via an Excel sheet. Matlab Simulink in general can be used to create and control technical, physical or mathematical systems.

5.1 FEMM

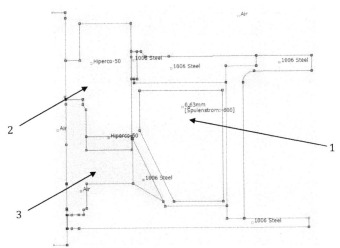

Figure 6 - The mashed solenoid in FEMM[14]

---

[14] Rommelfanger et al. (2014): Nachbildung des menschlichen Herzens

In figure 1 the solenoid is shown. A currency gets induced to the coil [1] and the anchor [2] is moving downwards, minimizing the airgap [3].
To get a dynamic model, every step has to be simulated: Showed are the steps with different airgaps.

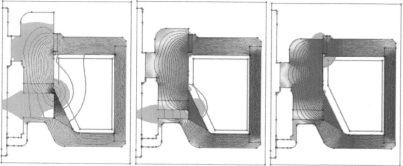

Figure 7 – Static FEMM models with 7, 2.5 and 0.1 mm airgaps (from l to r)[15]

Seen in Figure 7 is that the saturation of the solenoid it getting higher by minimizing the airgap.

---

[15] Rommelfanger et al. (2014): Nachbildung des menschlichen Herzens

5.2 Creating the Matlab Simulink model

The created Matlab Simulink model out of the predefined equations, shown in Figure 8, has to be filled with the calculated values. See Appendix F(delta, I) and F(delta, L). The model is divided in the two parts electric and mechanic, which are described in the following subchapters.

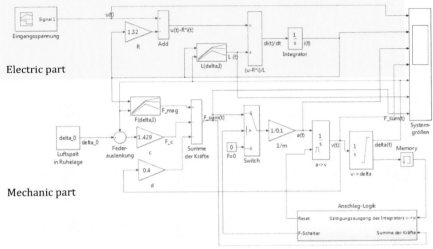

Figure 8 - Matlab Simulink model of the solenoid movement[16]

5.2.1 Electric part of the Simulink model

The voltage $U(t)$ is the only signal that can be changed manually.
The currency $I(t)$ can be calculated with the input signal $U(t)$. Having the currency being changed in a loop is a difference of the incoming $U(t)$, which chances by time, and the product of resistance R times the branched outgoing currency $I(t)$.[17]

$$U(t) - R * I(t) \qquad \text{Simulink eq. 1}$$

The next step to calculate the currency $I(t)$, is to calculate the quotient of the Simulink eq. 1 and the inductivity L. The inductivity L is plotted into an array out of FEMM.[18]

$$\frac{U(t) - R * I(t)}{L} \qquad \text{Simulink eq. 2}$$

---
[16] Self creation in Matlab Simulink
[17] Kallenbach et al. (2012): Elektromagnete; S.48f
[18] Kallenbach et al. (2012): Elektromagnete; S.48f

Due the inductivity L is dependent of the time t, the whole Simulink eq. 2 is now dependent of the time. The last step to get the currency $I(t)$ is to integrate the function over the time t.[19]

$$I(t) = \int \frac{U(t) - R*I(t)}{L} dt \qquad \text{Simulink eq. 3}$$

5.2.2 Mechanical part of the Simulink model

The equation (9) can be set as a differential equation of the second order. Second order means, that the frequency of the mass forces is double as frequency of the actuator.
To get a dependency between the variables of motion, the airgap s has be derivated once to get the velocity $\dot{s}$. Another derivation is the acceleration $\ddot{s}$. In the equation (9) this change is shown.[20]

$$F_M = m*\ddot{s} - F_{HP} - F_s \qquad \text{Simulink eq. 4}$$

A given condition is the inductivity L and the magnetic force F_M. These values are plotted into an array out of FEMM. A sequence of the current airgap s_{curr} will control the force F_M by reducing the distance.

With 'delta_0', the airgap $s = 7mm = 0.007m$, the mechanical loops starts. A difference of the airgap and the moved gap s_{curr} is multiplied by the spring constant c. The target is the spring force F_s:[21]

$$F_s = (s - s_{curr})*c \qquad \text{Simulink eq. 5}$$

The last input is the velocity $\dot{s}$, which is a derivation by time of the new current airgap $(s - s_{curr})$. Over the damping constant d, the system's damping force F_D can be calculated.[22]

$$F_D = (\dot{s} - \dot{s}_{curr})*d \qquad \text{Simulink eq. 6}$$

The now known forces are given as a summery in the following equation:

$$F_{Sum} = F_M - F_S - F_D \qquad \text{Simulink eq. 7}$$

---

[19] Kallenbach et al. (2012): Elektromagnete; S.48f
[20] Binder (2012): Elektrische Maschinen und Antriebe; S.11f
[21] Binder (2012): Elektrische Maschinen und Antriebe; S.11f
[22] Binder (2012): Elektrische Maschinen und Antriebe; S.11f

F_{Sum} is now devided by the anchor mass m, to get the anchor's current acceleration.

$$a_{curr} = \frac{F_{Sum}}{m}$$ Simulink eq. 8

The last steps are to integrate the acceleration to get the current velocity v_{curr} and then the current airgap s_{curr}.

$$v_{curr} = \int a_{curr}\, dt$$ Simulink eq. 9

$$s_{curr} = \int v_{curr}\, dt$$ Simulink eq. 10

6 Use of the system

The Simulink model used in chapter 5.2, has one input value: The voltage. This is a manually decreasing variable over the time. The voltage can be changed with a so called "signal builder".

The voltage of the input signal can be changed every $0{,}005s = 5ms$. This is a duty cycle of 200Hz. With this period of time for DC, the currency won't reach the value to have the force $F_S = 100N$. By extending the cycle time, the setting for the currency can be made. The Figure 9 shows a diagram at which the voltage level decreases over time until $0{,}05s$.

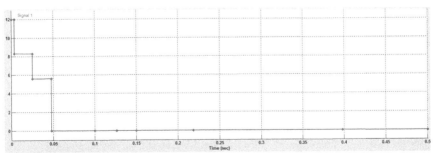

Figure 9 - Voltage level by time (num.1)[23]

With this voltage, the currency increases immediately, so the force rises up to $100N$ within a tick of time and the solenoid accelerates over the small airgap in $0{,}1$ seconds.

---

[23] Screenshot of Matlab Simulink voltage input

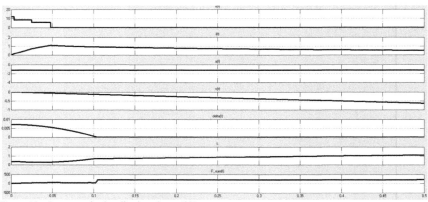

Figure 10 – Scope results for all values (num.1)[24]

As seen in Figure 10, the currency $i(t)$ increases in $t = 0,05s$ up to $i(t = 0,05s) = 1A$. In this period of time, the Force $F\_sum(t)$ increases just up to $F\_sum(t = 0,05s) = 50N$. First, the force is not high enough to get the system into movement. A force of about $100N$ is needed. Second, if the system moves, delta(t) - the airgap - is getting minimized within $t = 0,1s$. Third, if the system moves and the airgap can be minimized at $t = 0,1s$, the pressure which is produced, is enormous high for an idly system like the human blood circulation.

The airgap is moving like the PT1 signal, into the upper region of $delta(t) = 0,01mm$. Because the airgap is decreasing continuously and the force F_M is not, the acceleration of the anchor is constant, so the velocity is increasing and the time for the anchor to hit the HP-side is getting shorter.

7 Conclusion

The manual signal, which is used to control the voltage for a DC, is not the correct input for a solenoid using as an actuator. If the magnet would get the force to move the anchor, the pressure which goes into the HP-side would destroy the veins and arteries. The chance to get a turbulent flow in the chambers is very high, so the chance to get an apoplectic stroke is available.

An actuator like this direct current solenoid is not controllable over the time and the voltage.

---

[24] Screenshot of Matlab Simulink voltage output

A controllable solenoid like a proportional solenoid has to be used to control every movement the solenoid is doing or the system has to get higher active damping plus a higher friction of the solenoid with the surroundings to maximize the moving time.

All in all, the solenoid would work, unfortunately not how is currently. Some changes have to be made to optimize the result and have a continuous movement in HP and LP-side. We see the results as compliment of what the whole group has done this semester positive, the target to reduce the weight and to get a higher magnetic force has been achieved. The next step would be to add an active damping system so the anchor can move within a defined period of time. In there the damping force has to be spreaded into the active and the passive part, means the movement into the HP-side and LP-side, so that the force of the spring and magnet would have the same time for the complete movement of the anchor. The other idea is, as written before, to exchange the solenoid as a direct current engine to a proportional magnet, with which every step can be controlled. Problems hereby may be the complete reconstruction of the new solenoid and to repeat the analyses for the temperature and thermodynamic part and the complete new design of the heart.

8 Table of Figures

Figure 1 – Final situation of the solenoid from the group of winter term 13/14 3
Figure 2 - Initial situation at the beginning of the simulation and optimization 4
Figure 3: symbol figure for a PT1 unit step response 6
Figure 4 - Measurements of the coil 7
Figure 5 – MSDS of the solenoid 8
Figure 6 - The mashed solenoid in FEMM 11
Figure 7 – Static FEMM models with 7, 2.5 and 0.1 mm airgaps (from l to r) 12
Figure 8 - Matlab Simulink model of the solenoid movement 13
Figure 9 - Voltage level by time (num.1) 15
Figure 10 – Scope results for all values (num.1) 16
Figure 11 – F(delta, I) Measurements out of FEMM 20
Figure 12 – F(delta, L) Measurements out of FEMM 20

9 References

Bauer, Kerstin and Enneker, Juergen (2008): Herzklappenchirurgie – Ein Patientenratgeber; Steinkopff Verlag; Lahr.

Binder, Anreas (2012): Elektrische Maschinen und Antriebe – Grundlagen Betriebsverhalten; Springer Verlag; Heidelberg.

Boege Alfred and Boege Wolfgang (2013): Handbuch Maschinenbau – Grundlagen und Anwendungen der Maschinenbau-Technik; Springer-Vieweg; Wiesbaden.

Heidrich, Peter (2014): Magnetism – Fundamentals; Scriptum Electric Machines; HS Pforzheim.

Kallenbach, Eberhardt et al. (2012): Elektromagnete - Grundlagen, Berechnung, Entwurf und Anwendung; Springer Fachmedien; Wiesbaden.

Riewe et al. (2013): Vollständige Nachbildung des menschlichen Herzens - Gleichstromhubmagnet mit Federrückstellung; Konstruktionsseminar WS2013-2014; HS Pforzheim.

Rommelfanger et al. (2014): Nachbildung des menschlichen Herzens - Entwicklung und Optimierung eines Gleichstromhubmagneten mit Federrückstellung; Konstruktionsseminar SS2014; HS Pforzheim.

11 Appendix

11.1 FEMM Files

11.1.1 F(delta, I)

I [A] \ s [m]	0,00000	0,00010	0,00025	0,00050	0,00100	0,00150	0,00200	0,00250	0,00300	0,00350	0,00400	0,00450	0,00500	0,00550	0,00600	0,00650	0,00700
0,01000	1,00552	0,15779	0,02740	0,01170	0,00753	0,00621	0,00573	0,00557	0,00567	0,00576	0,00595	0,00620	0,00647	0,00675	0,00709	0,00742	0,00784
0,16200	164,94600	39,49590	7,25045	3,09479	1,99878	1,64637	1,51937	1,48060	1,50411	1,52987	1,58116	1,64514	1,71551	1,79302	1,88078	1,97274	2,07847
0,32400	218,78600	91,96210	27,65330	12,13290	7,93909	6,57380	6,10687	5,98989	6,10782	6,25305	6,48144	6,76181	7,05892	7,37394	7,69310	7,99184	8,24794
0,48600	222,78400	112,39000	42,74800	24,13010	18,56700	15,77600	14,70210	14,34310	14,39830	14,52750	14,74010	15,00360	15,26220	15,50230	15,72830	15,88100	15,89450
0,64800	219,31500	115,52000	47,13730	35,22270	32,31130	28,12540	25,82720	24,65740	24,20540	23,97980	23,91930	23,97200	24,04480	24,11080	24,15630	24,11500	23,86360
0,81000	215,42600	116,01500	48,88760	43,15040	45,58120	41,09490	37,86000	35,77830	34,64350	33,93420	33,51100	33,29080	33,12930	32,98280	32,83580	32,56740	32,04440
0,97200	211,75800	116,11700	50,57750	49,36090	56,68940	53,64460	50,20470	47,44710	45,60820	44,31870	43,45880	42,90460	42,46620	42,08720	41,71510	41,19620	40,39490
1,13400	208,43900	116,25100	52,31070	54,31920	66,55920	64,79190	62,34570	59,40560	56,99160	55,10480	53,74340	52,79790	52,04070	51,39660	50,76450	49,97220	48,86880
1,29600	205,51700	116,53300	54,35200	58,47610	75,38190	74,76410	73,72740	71,28160	68,63840	66,22330	64,33370	62,95010	61,83680	60,89030	59,97050	58,88430	57,46170
1,45800	202,89900	116,96400	56,78950	62,13500	83,18680	83,95910	83,88550	82,79940	80,31970	77,57680	75,19790	73,34170	71,83340	70,55180	69,32210	67,92330	66,15640
1,62000	200,50100	117,53100	59,47340	65,55600	90,10920	92,39440	93,35300	93,40750	91,82300	89,05880	86,28310	83,96040	82,02570	80,37650	78,81050	77,07970	74,95140

F_M [N]

Figure 11 – F(delta, I) Measurements out of FEMM

11.1.2 F(delta, L)

I [A] \ s [m]	0	1,00E-04	2,50E-04	5,00E-04	1,00E-03	1,50E-03	2,00E-03	2,50E-03	3,00E-03	3,50E-03	4,00E-03	4,50E-03	5,00E-03	5,50E-03	6,00E-03	6,50E-03	7,00E-03
0,01	2,380820	1,591050	1,278530	1,191370	1,108300	1,040980	0,981505	0,924940	0,868410	0,811033	0,752165	0,691075	0,6276	0,561549	0,491719	0,419073	0,34192
0,162	1,950050	1,585450	1,288850	1,201220	1,117380	1,049160	0,988857	0,931484	0,874090	0,815782	0,755861	0,69372	0,629128	0,56189	0,490881	0,41682	0,338608
0,324	1,45798	1,386080	1,256090	1,177960	1,096520	1,028930	0,968033	0,908760	0,848511	0,78629	0,721241	0,653914	0,584301	0,513817	0,441663	0,369515	0,298094
0,486	1,102650	1,097160	1,076900	1,043890	0,963044	0,885923	0,817031	0,753521	0,692937	0,634266	0,577059	0,521015	0,466057	0,412065	0,359003	0,307014	0,256457
0,648	0,873785	0,874301	0,872627	0,858547	0,798747	0,734640	0,676098	0,622764	0,573098	0,52589	0,480439	0,436294	0,393268	0,351213	0,309999	0,269724	0,23059
0,81	0,726242	0,727233	0,727084	0,718997	0,680404	0,631575	0,583459	0,538263	0,496117	0,456379	0,418409	0,381755	0,346182	0,311472	0,277639	0,244629	0,212597
0,972	0,624278	0,625168	0,624544	0,618972	0,591783	0,556468	0,517445	0,478916	0,442316	0,407756	0,374912	0,343369	0,312874	0,283241	0,254415	0,22634	0,199159
1,134	0,548916	0,549631	0,548530	0,544123	0,524421	0,497410	0,466888	0,434267	0,402186	0,371567	0,342473	0,314645	0,287837	0,261877	0,23668	0,212195	0,188531
1,296	0,490585	0,491141	0,489743	0,486032	0,471436	0,449892	0,425958	0,398780	0,370763	0,343383	0,317209	0,292216	0,268212	0,245037	0,22259	0,200827	0,179843
1,458	0,444102	0,444536	0,442948	0,439683	0,428463	0,411133	0,391415	0,369428	0,345102	0,320600	0,296855	0,274109	0,252311	0,231327	0,211049	0,191424	0,172533
1,62	0,406251	0,406586	0,404885	0,401932	0,392872	0,378772	0,362355	0,344049	0,323443	0,301618	0,279988	0,259121	0,239116	0,219897	0,20136	0,183462	0,166267

L [Vs/A]

Figure 12 – F(delta, L) Measurements out of FEMM

11.2 Coil Calculations

At first the number of windings in the blue are will be calculated. Taken from the geometry, the following equations are being used:

$$r_1 = r_a - r_i \qquad \text{eq. 10-011}$$

$$n_{r_1} = \frac{r_1}{d} \qquad \text{eq. 10-012}$$

$$z_1 = l - l_s \qquad \text{eq. 10-013}$$

$$n_{z_1} = \frac{z_1}{d} \qquad \text{eq. 10-014}$$

$$n_b = n_{r_1} \times n_{z_1} \qquad \text{eq. 10-015}$$

Secondly the same is will be calculated for the red area:

$$r_1 = r_a - (\frac{r_s - r_i}{2} + r_i) \qquad \text{eq. 10-016}$$

$$n_{r_2} = \frac{r_2}{d} \qquad \text{eq. 10-017}$$

$$z_2 = l_s \qquad \text{eq. 10-018}$$

$$n_{z_2} = \frac{z_2}{d} \qquad \text{eq. 10-019}$$

$$n_r = n_{r_2} \times n_{z_2} \qquad \text{eq. 10-10}$$

Finally, there is the addition of both windings. To get the optimum winding number, an 85% tolerance will be added to the final amount. This percentage is an empiric value.

$$n = (n_b + n_r) \times 0.85 \qquad \text{eq. 10-11}$$

The middle radius of the coil for both areas are calculated with the following formulas:

$$r_{m_b} = \frac{r_a - r_i}{2} + r_i \qquad \text{eq. 10-12}$$

$$l_b = 2 \times \pi \times r_{m_b} \times n_b \qquad \text{eq. 10-13}$$

$$r_{m_r} = \left(\frac{r_a - \left(\frac{r_s - r_i}{2} + r_i \right)}{2} \right) + r_i \qquad \text{eq. 10-14}$$

$$l_r = 2 \times \pi \times r_{m_r} \times n_r \qquad \text{eq. 10-15}$$

Now we sum the lengths of both areas, to calculate the total length of the wire for the entire coil. Experiences have shown that the winding is made to 85% optimally.

$$l = (l_b + l_r) \times 0.85 \qquad \text{eq. 10-16}$$

The results of the calculations are in the following Table. All values are in mm.

Number of Windings blue		Total Number of Windings	
r_1	14.6	n	549.7
n_{r1}	23.17		
z_1	7.15	**Average Coil radius**	
n_{z1}	11.35	r_{mb}	22.7
n_b	263	r_{mr}	29.3
Number of Windings red		Lengths of Areas	
r_2	11.85	l_b	37508.4
n_{r2}	18.81	l_r	69990.9
z_2	12.85		
n_{z2}	20.4	**Total Length**	
n_r	383.72	l	91371

YOUR KNOWLEDGE HAS VALUE

- We will publish your bachelor's and master's thesis, essays and papers

- Your own eBook and book - sold worldwide in all relevant shops

- Earn money with each sale

Upload your text at www.GRIN.com
and publish for free